Harvesting
Honey

by Wally Shaw

Northern Bee Books

Harvesting Honey
© Wally Shaw

ISBN 978-1-912271-36-8

Published by Northern Bee Books, 2018
Scout Bottom Farm
Mytholmroyd
Hebden Bridge HX7 5JS (UK)

With the agreement of Cymdeithas Gwenynwr Cymru
The Welsh Beekeepers Association.Bridge HX7 5JS (UK)

Design and artwork by DM Design and Print

Printed by Lightning Source UK

Harvesting
Honey

by Wally Shaw

Contents

Harvesting Honey

Introduction

The aim of this booklet is to help beekeepers to better understand honey itself and to harvest and prepare it for home use or sale retaining as much of its essential properties as possible. What exactly is honey, for it is certainly a lot more than a solution of various sugars in water? If we are to produce good honey it is important to understand how it should be handled in all stages between the hive and jar because in reality it is quite a delicate product. Stories about finding four thousand year old honey in Egyptian tombs and "and it was just as good as the day it was put there" are just that – stories. There are many similarities between honey and wines; they both need great care in their production, handing and storage if they are to develop and retain their full potential.

It is generally accepted that the `best` honey comes straight from the comb (cut comb or sections). The various processes that we use to get pristine honey from the comb into a jar all have the potential to damage it in some way. In Britain we currently get a premium price for home-produced honey (the envy of beekeepers in many other parts of the world) and it is our responsibility to see that we bring a top quality product to market.

What is Honey?

Floral honey comes from flowers in the form of nectar which is a sugar solution. Nectar is an inducement evolved by flowers to encourage insects to visit them and effect pollination (ie. transfer pollen from flower to flower). A few plants also have what are called extra-floral nectaries, eg. Cherry-laurel, where a sugary solution is produced for some other purpose. Often this is to encourage ants (or some other predatory insect) to live or hunt on them and give them protection against plant eating insects. There is a more primitive use of nectar because some non-flowering plants, such as bracken (ferns evolved long before flowering plants), have extra-floral nectaries. Another source of sugar (it can not really be called nectar) is honeydew which is produced by sap-sucking insects. Plant sap is rich in carbohydrates (sugars) but has relatively low levels of proteins, so in order to obtain a balanced diet, such insects have to excrete the excess. The sugars excreted this way are mostly metabolites (altered sugars) and, because they and other constituents are so different from those found in nectar, some authorities consider honeydew honeys not to be true honey.

When nectar arrives at the hive it has already had three enzymes added to it. These are all products of the hypopharyngeal gland. Invertase (now usually called sucrase) is the only one concerned with honey processing per se and its function is to split the di-saccharide sucrose into the mono-saccharides fructose (laevulose) and glucose (dextrose). These are the two main sugars found in honey and occur in roughly equal proportions (average 38% and 31% respectively). The ratio of these sugars is quite variable and high fructose honeys will remain liquid (clear) whereas high glucose honeys will granulate (set). The only other sugar that is present in floral honey in any significant amount is maltose (average about 7%). What is often referred to as ripening of honey is the combination of drying (the removal of water by evaporation) and inversion of sucrose. Some plants produce nectars that are particularly rich in sucrose; borage is the most well known example but lavender nectar is even higher. The amount of sucrase that the bees add to the nectar does not appear to be a response to the types of sugar present but more to do with the physiological condition of the bees and the time of year when collection occurs.

The two other enzymes the bees add to honey are diastase and glucose oxidase. Diastase breaks down starch and is probably used in the digestion of pollen (nothing to do with honey). Glucose oxidase is important because it breaks down glucose to produce gluconic acid and hydrogen peroxide. The latter has anti-bacterial activity that helps preserve the honey and is also part of the medicinal properties of honey. The acidity of honey (average pH 3.9) also contributes to its preservation.

Combined solutions of fructose and glucose have some peculiar properties without which honey as we know it would not exist. The main property is that a strong solution of fructose will dissolve more glucose than a weak one. Even more peculiarly, this high combined solubility of fructose and glucose only operates over a limited temperature range which just so happens to include the temperature at which honey is processed inside the hive. The net result is a highly concentrated solution of sugars which could not be produced at a lower temperature of 10-20°C. This supersaturated solution provides the colony with a compact energy source which, because of its high osmotic pressure (and anti-microbial properties), is not subject to degradation due to yeasts (fermentation) or bacteria.

However, in order to be free from the risk of fermentation, the water content of honey has to be reduced to below 18.5%, which is sometimes a problem for colonies operating in a humid oceanic climate. Although sealed (capped) honey is said have a guaranteed low water content this does not always prove to be the case. It may have been low at the time of sealing but in a wet season it is not uncommon for honey to pick up water through the slightly permeable cappings, especially if combs are left on the hive as the number of bees is rapidly diminishing. High water content (sealed) honey can also occur in hives where the number of bees has been radically reduced by swarming. If it is on the borderline for water content, honey that granulates is more prone to fermentation. This is because water expelled during crystallisation causes honey in the interstices (between the crystals) to have a higher water content which may allow yeasts to become active. Honey sets most rapidly in the temperature range of 10-18°C.

What Else Does Honey Contain?

According to Eva Crane ('A Book of Honey') 181 different substances have been identified in honey, some of which are not known to exist elsewhere. This book was written over 30 years ago and the technology of chemical analysis has advanced considerably since that time so goodness knows what the tally is now. Honey contains no fat and only a very small quantity of protein, probably in the form of pollen grains. It also contains a range of vitamins and trace minerals but none of these occur in quantities that make a significant contribution to human dietary needs. It is not known how important these substances are for the bees. Small amounts of compounds that are thought to act as anti-oxidants are also present but again these seem to have very little dietary significance for us humans.

The enzymes added by the bees also have no direct dietary significance nor do they contribute directly to the flavour. Enzyme activity in honey is often mentioned because it can be used to assess the quality of the honey or, more precisely, the degree of damage that has been inflicted on it during processing. This is something that is relatively easy to measure and a low level of enzyme activity implies that honey has been over-heated.

For the consumer it is the flavour of honey that really matters and most of the flavours that we so value are produced by the plants from which the nectar was gathered. All we really need to know about flavours (and most of which have not been identified in a chemical sense) is that they are volatile. This means that exposure to air, particularly at elevated temperatures, results in a progressive loss of flavour. Smell and taste are inextricably mixed in our sensory perception of honey so it seems almost inevitable that if you can smell honey it is in the process of losing something. The precautions to prevent this loss are therefore obvious; minimum temperature and limited exposure to air. This is why comb honey has the best flavour.

Hydroxymethylfurfural (HMF)

Often talked about but little understood, HMF arises through the dehydration of fructose. The rate of production of HMF increases exponentially with temperature (4-5 times faster for each 10°C rise in temperature). It is also faster under acidic (low pH) conditions. To be legal for sale, the HMF content of honey must be less then 40mg per kg but there is a special dispensation for tropical honeys of 80mg/kg.

Testing honey for HMF was originally introduced to detect counterfeit honey which was usually based on industrially produced corn syrup. Until enzyme inversion came on the scene, corn syrups had a high level of HMF (the result of heat and acidic conditions during production) which gave the game away. Only later did the measurement of HMF become a means of assessing whether genuine honey had been subjected to excess heat during processing after it had left the hive. The amount of HMF does constitute quite a good yardstick for honey quality but it is still possible to damage honey more than is necessary without straying beyond the legal limit.

At the levels we encounter it in our diet, HMF is not toxic to humans; so the legal requirement for the level of HMF in honey is certainly not a health (food safety) issue. Jams, sugar syrups, molasses and many soft drinks all have HMF levels that are 10-100 times greater than those in honey and all cooking of foods that contain sugar (which is virtually everything really) produces some HMF. HMF is a by-product of caramelisation, so all cakes and biscuits contain high levels.

HMF is toxic to bees and, in experiments at Rothampstead many years ago, 7 year old honey was found to kill bees (the level of HMF and the storage history of the honey were not specified). So the message here is, don't feed your bees that old tub of honey that you have just found at the back of your shed!

Honey straight from the hive typically has less than 15mg/kg of HMF but this will inevitably increase during processing and during storage prior to consumption. The aim should be to keep this increase to a minimum before it is in the jar ready for sale. After that it is up to the consumer (guided by the Best Before date you have put on the label) to look after the honey.

Heating Honey

In an industrial setting (a honey bottling plant) honey is often flash-heated to 60-70°C for ease of filtering; it is rapidly heated, held at the higher temperature for no more then 5 minutes and then rapidly cooled. This regime minimises the production of HMF but what does it do to the quality of the honey? It does pasteurise the honey and kills most of the yeasts that might cause fermentation (although some of the *Zygosaccharomyces* species are quite heat resistant and may survive). Pasteurisation is, of course, totally unnecessary if the honey has a low water content. Exposure to a temperature in excess of 60°C also delays granulation and ensures a good shelf-life as clear honey (but at the price of flavour). The downside is that when granulation does occur it is often patchy, layered and not visually pleasing. Perhaps fortunately, the equipment required to flash-heat honey is beyond the resources of the ordinary beekeeper so we are reduced to more low-tech (and potentially less damaging) methods.

On the basis that whilst in the hive honey is never exposed to temperatures in excess of 37°C it is recommended that during processing by the beekeeper heating should be limited to 40°C. Even then, this (maximum) temperature should only be used to liquefy honey that has set in the bucket during storage. The secret of heating honey without inflicting damage is to go slowly but the rate of heating and the time taken need to be carefully balanced. Like water, honey is a very poor conductor of heat but, unlike water, honey is viscous (or solid when set) and convection contributes little to the transfer of heat. Poor transfer of heat means that the temperature gradient used during heating has to be very gradual if local over-heating is to be avoided. For example, heating honey in a water bath with the water temperature set at 60°C sounds gentle enough but, without constant stirring, the layer of honey adjacent to the walls of the container will remain at close to 60°C for several hours before the bulk of the honey is up to temperature and this is undesirable. It is true that this layer of 'damaged' honey will be mixed with the rest of the honey later in the process, and will probably be barely noticeable, but why let this happen if it can be avoided?

The best way to warm honey is in a thermostatically controlled warming cabinet. These can be purchased from beekeeping equipment suppliers but an old refrigerator body (decommissioned by the removal of the refrigerant) is a cheaper option (see section on **Preparing Honey for Bottling** for details).

Honey Foraging

The nectar offered by plants varies widely in its sugar content, the amount produced and its ease of collection. Nectar from species such as acacia (*Robinia pseudoacacia*) is very concentrated with over 60% sugar (40% water). Most nectars seem to fall in the range of 20-45% sugar but what is really notable is how widely this can vary over time for some species, eg. blackberry 15-45%. Presumably this variation is driven by weather conditions (temperature, relative humidity and wind) and by the water relations of the plant (soil moisture and transpiration rate).

As beekeepers we can do absolutely nothing to influence from what plants our bees collect except by moving the hives to different foraging areas. The colony's strategy for nectar collection is fairly well understood and is nothing to do with collecting the best flavoured nectars. Basically it is a matter of energy budget; most energy source (sugar) collected for the least expenditure of energy. The final act of the foraging dance routine is for the dancer to offer her 'audience' a free sample so that they can assess the quality of the product (the sugar content). This also provides the recruited foragers with some smell information to help them locate the source. Apart from sugar content, other factors that influence the choice of a nectar source are flying distance, the number of flowers that have to be visited and the time taken to collect a full load of nectar.

As a result of the up-to-date information provided by the incoming bees that perform on the dance-floor and interactions between incoming foragers and receiver bees (the ones that store the nectar in the cells), the colony is able to achieve continuous optimisation of its nectar collection. Usually sugar-rich nectars are preferred over dilute nectars but there are exceptions. In spring, when there are still stores that need to be diluted for use, weak nectar may be preferred to something stronger. Again, in hot weather, nectar with a high water content may be preferred for its value in cooling the hive (evaporative cooling). It is the receiver bees (who have knowledge of the in-hive conditions) who modify the foraging strategy by unloading bees that are carrying what they most urgently need more quickly. A bee that is unloaded quickly responds by going straight out to get more of the same, whereas a bee that suffers a delay in unloading is less enthusiastic and may change what it is collecting. A remarkable control system that is simple but effective.

Harvesting Honey

Taking Advantage of Different Honeys

Honeys produced in different places and at different times of the year can be quite variable, offering a range of flavours. There are many who say that what the market wants is a uniform product – the 'supermarket mentality'. However, the experience of many beekeepers is entirely different and that the purchaser of local honey is much more discerning and enjoys the taste of different honeys.

Yes, it is bit of a pain extracting honey at different times of the season, as the opportunity presents itself, and most beekeepers prefer to do it at the end of the season in a single, epic (messy) session. If you only want to do one extraction, an alternative solution is to number the supers as they go on the hive and then extract all the number 1's together, all the number 2's and so on. Keeping the different flows and different apiaries separate can enhance the saleability of your honey. It is not something commercial beekeepers can do but for hobby beekeepers it is an option, if you are prepared to take the trouble.

Figure 1

Another advantage of taking intermediate crops of honey during the season is that it economises on equipment – supers can be extracted and the boxes and combs returned to the hive to be re-filled. It also reduces the amount of lifting that is required to make routine inspections of the brood area for the rest of the season. However, beware of going to extremes with early removal because subsequent bad weather could lead to colony starvation. If oil-seed rape honey is involved you have no choice but to take an intermediate harvest as soon as flowering is in decline. If the bees have access to autumn sown oil-seed rape then this honey will almost certainly have to be harvested in April or early May. Extraction must also be completed without delay or the honey will set in the combs.

Which combs can I take?

In an ideal world honey would come in whole boxes of completely sealed combs and, indeed, this may be substantially the case at the end of the season in a good summer. But in the real world, and particularly in the case of intermediate harvests, the beekeeper is faced with incomplete boxes of combs with a variable proportion of sealing. Vigorous shaking of combs with unsealed contents will reveal whether newly collected nectar is present (droplets will spray out of the shaken comb). This allied with a knowledge of local nectar flows, recent activity of the colony (has it been foraging hard) and the weather will also help you decide which honey is safe to take. The ultimate test is to use a refractometer to measure the water content of the honey in unsealed cells; a dip with a matchstick in several cells will provide a large enough sample to take a measurement. Do not just sample combs in the middle of a box as they will almost invariably show a lower water content that those towards the outside. This is because of the way air circulates in the hive; up the middle and down the sides. It is always best to err on the safe side because once honey has actually been extracted it is virtually impossible to remove any excess water - blending with low water content honey is the only option.

However, it is quite easy to apply supplementary drying after the honey has left the hive if it is still in the combs because these have a large (vertical) surface area for gaseous exchange – this is how the later stage of drying occurs in the hive. If you are beekeeping on a scale that justifies having a well–sealed honey super storage facility and a dehumidifier the problem is easily solved. Small, low capacity dehumidifiers with a power consumption of about 200W can be purchased for just under £100 and will quickly dry honey that has a high water content. A dehumidifier will easily maintain a relative humidity (RH) below 50% in a sealed room. Good sealing of the room is important because you want to minimise incoming air from which water will have to be removed. The equilibrium water content of honey at an RH of 50% is 15.9%, which is probably lower than can be achieved direct from the hive even in the driest part of Britain. In practice equilibrium would take several

Figure 2

weeks to achieve but 5-7 days of drying with good air circulation round the boxes of combs is sufficient to reduce the water content from 19-22% down to 17.0-17.5%. Temperature has a major influence on the rate of drying and something around 25°C is ideal. The yield of water in the dehumidifier is a good guide as to progress because it falls off rapidly as the drying process is nearing completion. In financial terms it will only take the drying of about 30lbs of honey to recover the outlay on the dehumidifier, a piece of equipment that should last for years.

Getting honey off the Hive

There are a number of ways honey can be removed from a hive; some are quite benign and gentle and others nothing short of barbarous and have the potential to damage the colony and/or contaminate the honey. Here are the main methods in reverse order as to their desirability.

- Chemical repellents – Originally a cloth impregnated with carbolic acid was used but has long been banned for several good reasons. More recently benzaldehyde has been used on a board or cloth and I am not sure about the current legality of this chemical. There are other products on the market containing what are claimed to be natural or harmless repellents but I have no idea what the active ingredients are. Some beekeepers who have used them say that they will never do so again!

- Blowing bees from the supers – This is accomplished using a mechanical blower (often a leaf blower). This is not really a method for the hobby beekeeper but some bee farmers use it. Because hive bees (those that have never been outside the hive before) will be blown into the air or onto the ground this method could never be described as sympathetic.

- Shaking (or brushing) bees from individual combs - If the bees are shaken back into the hive, rather than into the air or onto the ground, this is not a bad method for removing a small number of combs. It is useful for taking an intermediate crop of honey where just some of the combs in a box are being removed; the required combs can be shaken (or brushed) clear of bees and immediately replaced with empty drawn frames or foundation with little hassle. It is important to have an empty box available to receive the combs and a cover to keep bees out. Unless done discretely and fairly quickly under the right conditions (when there is some sort of a nectar flow) shaking frames can promote robbing.

- Using clearer boards – This is by far the most common method of removing bees from supers so that the honey can be harvested in a more genteel and less disruptive (to the bees) manner. Use of clearer boards is particularly suited to taking the main honey crop at the end of the summer. Most cover boards double up as clearer boards and have two holes in them that fit the standard Porter escape. Porter escapes work really well provided the beekeeper pays attention to the detail. The springs in the escape must be correctly adjusted and be free from propolis. The escapes should be pinned or taped to the board so that they can not become dislodged during use. The practice of storing Porter escapes in the cover board when they are not in use is to be deplored; it ensures the escape will not work properly when required and also blocks off top ventilation to the hive. There are several other designs of clearer board on the market; I have no experience of their use but if they have stood the test of time I am sure they work well enough. When using Porter escapes it helps to have extra space between the bottom bars of the frames and the board. The mini-eke that many beekeepers use during Varroa control is ideal for this purpose. It should be noted that most of the 'alternative' clearer boards incorporate such a space in their design. A cool night during which the bees want to move down into the warmth of the brood area also makes for a good clearance.

A warning about the use of clearer boards

It is essential that the part of the hive above the clearer board (the supers and the roof) have no holes or cracks through which bees can pass. As soon as the supers have been vacated any such holes will be unguarded and robber bees will very quickly discover that there are 'goodies' within just for the taking. Ensuring there is no unauthorised access to your supers is even more important if you are going to leave the clearer boards in place for longer than overnight. The ideal way of deploying clearer boards is to put them on in the evening (when there is less chance of starting a robbing spree through the smell of exposed honey) and then to take the honey off fairly early next day before the bees have worked out how to get back up again.

Essential Precautions when Harvesting Honey

There are two important questions that the beekeeper needs to ask when taking a major crop of honey from hives (particularly that at the end of the season):-

1)   When the supers have been removed is there sufficient room left to accommodate all the bees? Sometimes it is necessary to add a box of empty combs to provide sufficient space until the colony gets smaller in readiness for winter. It can usually be removed 2-3 weeks later.

2)   Are there sufficient stores remaining to prevent the colony from starving before it can be fed? Honey is typically harvested when there is no significant nectar flow and if it has been a poor later summer the bees may not have been able to store honey around the diminishing brood area, ie. they may have virtually no stores when you have removed the supers. This situation can be dealt with by starting to feed immediately – within the next couple of days may be necessary!

Other tips about harvesting honey

•   All hives that are having honey removed from them (in fact all colonies in the apiary) should have entrance blocks installed prior to the great event. This will greatly reduce the chance of setting-up robbing.

•   If the honey supers have not cleared properly you should ask the question, why? It could be because the clearing board is not working properly but it could also be because a certain royal personage has found her way into your supers. Check before you start shaking bees off frames in your annoyance.

•   It makes for less sticky transport of supers if all the boxes are separated at the time the clearer board is installed. This will break and brace comb between the boxes and the bees, being the tidy little things they are, will clean-up any spilt honey before they vacate the box.

- The use of castelations in honey supers also contributes to clean transport; the frames are held firm and do not swing or rub together thus damaging the cappings.

- Removed supers should be stored securely where bees and wasps are unable to gain access. If they are not to be extracted immediately the storage space should be dry (low humidity) and preferably warm (see discussion above in Which combs should I take?)

A Final Warning before Extraction

Do not extract honey from frames unless you are sure the water content is sufficiently low. It is possible to further dry honey whilst it is still in the comb but virtually impossible to do anything after it has been extracted. Capped honey is normally safe to extract but uncapped always needs investigation preferably using a refractometer.

The practice of leaving (or returning) uncapped or partly capped honey on the hive for capping to be completed will only work if there is a substantial late nectar flow. In the absence of a significant flow the colony will use what is coming in as 'running expenses'. Honey in supers will gradually be moved down into the contracting brood area where it will provide easily accessed winter stores. Uncapped honey will be removed first but if the colony is short of stores in the brood area and there is no flow this will be followed by capped honey. This is the colony's natural behaviour to prepare for the oncoming winter. It may save you money on winter feeding but, since honey is much more valuable than sugar syrup, this is usually not a good bargain for the beekeeper. Also, if you leave uncapped honey on the hive into mid-September it may be supplemented by a nectar flow from ivy and that will not enhance the flavour of the honey (according to most people's perception).

Equipment for the Processing of Honey

All the equipment that is used in the processing of honey (extractor, ripener, storage buckets, sieves, etc.) must be made of stainless steel or food grade plastic. If honey is to be sold it is illegal for it to come into contact with any other material (except wood of course). Particularly to be avoided (for your own safety as well as legality) is any equipment with soldered joints because, as has already been noted, honey is quite acidic. The continued use of some of the older galvanised extractors and ripeners is no longer acceptable – and has been illegal for many years.

Uncapping Combs

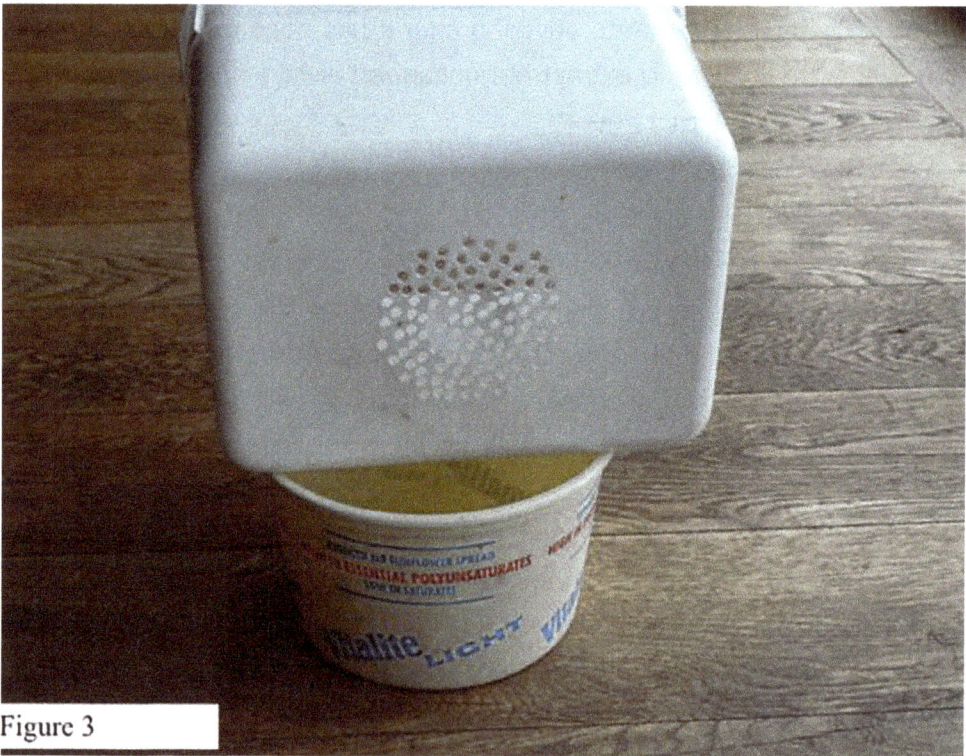

Figure 3

Nothing more complicated (or costly) than a ham or salmon knife with a blade length of 10-12 inches (25-31mm) is required to uncap honey. A sharp knife works perfectly well cold and there is no need to keep dipping it in a jug of hot water – as is sometimes suggested. Most people use the wooden surrounds of the frames as guide for the uncapping knife but if the combs are fat it is possible to cut higher and remove just the cappings. An uncapping fork or ordinary domestic fork can be used to uncap any surfaces that are below the level of the knife cut.

Uncapping Tray

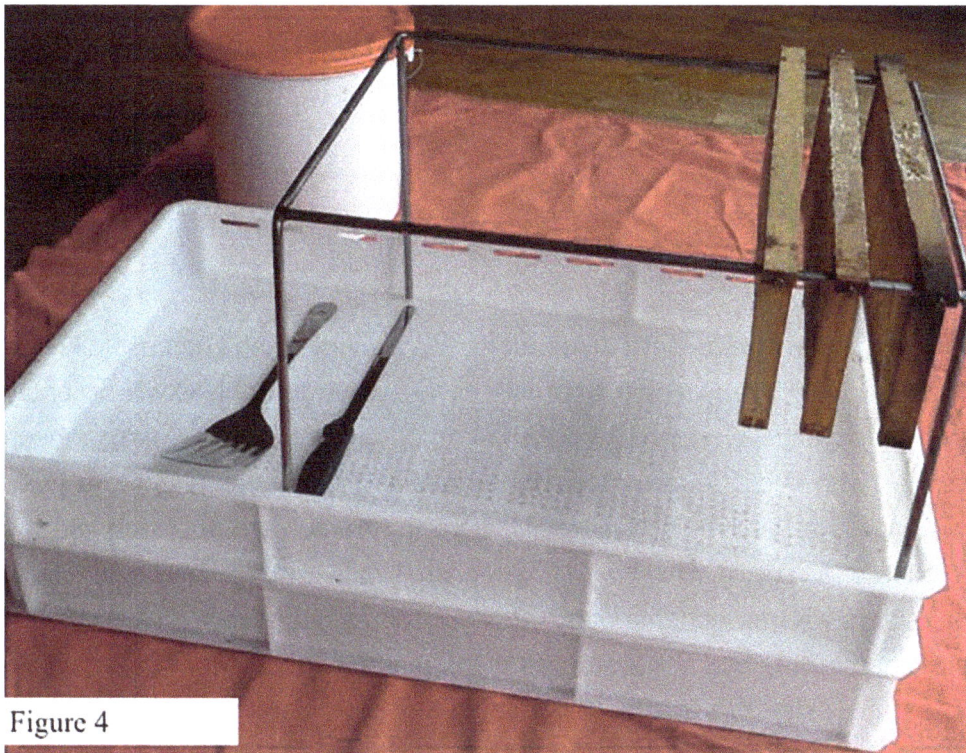

Figure 4

An uncapping tray is an absolute essential to keep the sticky business of extracting honey under control. A simple, unheated, plastic uncapping tray of the type shown is all that is required. A layer of cappings left on the perforated upper tray for a few hours will release most of its honey into the reservoir below. We then transfer the cappings into ice-cream boxes (2 litre size) which have had a circle of holes drilled in the bottom. These are placed over margarine tubs and put in a warm place and still more honey drains down. Finally the cappings can be warmed and spun in a perforated basket that replaces the cage on the shaft of the extractor. Long spin times (30 minutes or so) are needed to recover this honey which only constitutes about

1% of the total crop. I am not sure this final stage is really worth the effort and you might be better off leaching the cappings to make mead?

One thing not to do with cappings (I know because I have tried it) is to put them in a tray at the top of the hive for the bees to clean-up. Sounds like a good idea doesn't it? Yes, they do remove the residual honey but the 'poor dears' then regard the wax as badly damaged comb and set about trying to repair it. The result is a labyrinthine heap of wax firmly attached to the tray and with bees working away inside it. Sorting that out without killing bees is no mean task!

Extractors

For those who already have an extractor there is not much I can usefully say. Most beginners and people who only have few (up to 6) hives will probably have a tangential extractor (where the frames revolve, supported by a mesh screen, with one side facing out and one facing in,). Those with more hives will usually have a radial extractor, where the frames are arranged like the spokes of a bicycle wheel with the top-bars facing out. The smaller radial extractors will only accommodate shallow frames but usually have (optional) screens so they can double-up as a tangential extractor for large frames. Tangential extractors are usually cheaper and more compact (easier to store when not in use) but extract fewer frames (2, 3 or 4) at a time. The frames also have to be reversed at least twice during extraction, so the whole process of extraction is slower. Radial extractors are more expensive and have a minimum diameter of about 20 inches (508mm) so they take up more storage space when not in use. They can extract 9-12 frames at a time (commercial ones 24-48) and the frames do not need to be turned. Tangential extractors are usually manual (you turn the handle) and radial extractors can be either manual or motorised.

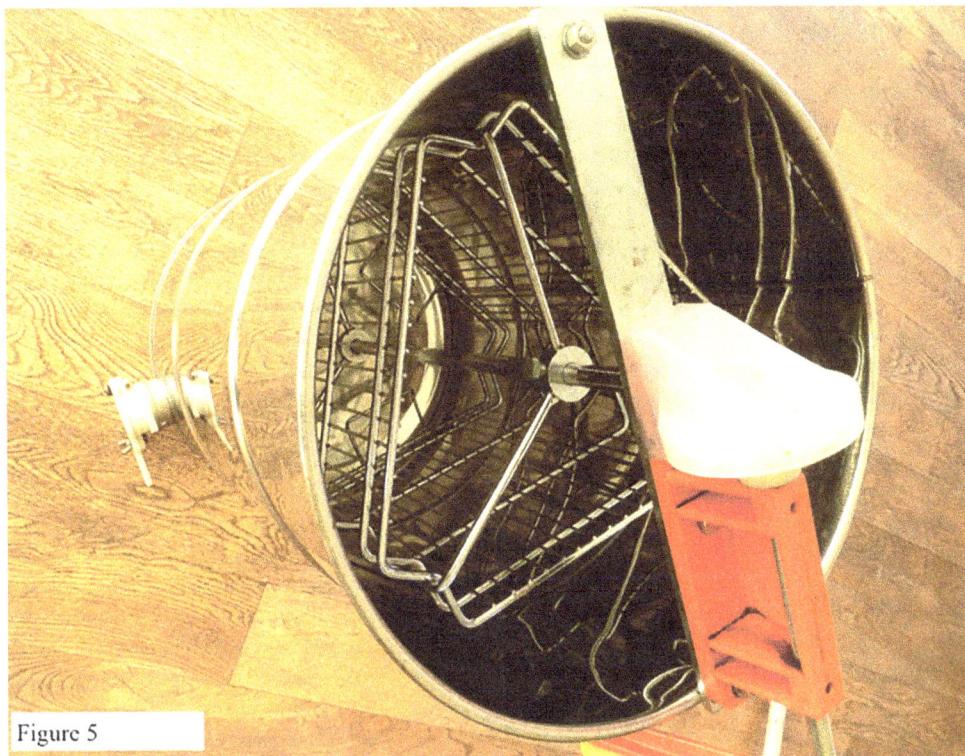

Figure 5

You have to get to know your own extractor; how best to load it, how fast you need to turn the handle or control the motor speed but here are a few tips about extraction.

- The extraction room and all working surfaces should be cleaned and warm water and clean cloths should be available to wipe up any spills. Clean polythene sheeting (500 gauge) over the working surfaces is something we employ. Obviously the room should be secure against the entry of wasps and bees. There should be no pets around and you should not smoke.

- Warm water and clean cloths should be available to wipe hands and clean up any spills.

- Make sure the extractor has been thoroughly cleaned before use and that the bearings are properly (but sparingly) lubricated. Petroleum jelly ('Vaseline') is best but sintered bronze ('Oilite') bearings may require a few drops of light oil with any surplus wiped off.

- A frame temperature of 20-25°C is ideal for extraction. A lower temperature will require a longer spin time and a higher temperature may result in comb damage.

- Try and achieve a balanced load and do not persist if the extractor is leaping about all over the place.

- The practice of screwing the extractor legs (if it has got them) down to a heavy base may seem like a good idea but it does encourage you to abuse the extractor more than is good for it.

- Start the extractor spinning slowly and gradually increase the speed as the weight of the combs decreases.

- Keep the extractor under supervision until you are sure that a good balance is going to be achieved (don't just walk away and leave it!).

- Remember that frames of free-flowing honey will automatically come into balance as the speed of rotation is gradually increased. If they don't there is something there in a frame (or frames) that is preventing this happening (usually a patch of granulation or pollen).

- Set such frames to one side for extraction as a separate load which can be spun more slowly for a longer time. It may be necessary to stop the extractor and rebalance the load as the frames are gradually relieved of their contents.

- Above all be careful because a fast spinning extractor can do an awful lot of damage to stray fingers.

Figure 6

Heather Honey

Extracting heather honey is quite a specialised subject that cannot be covered in detail in this booklet. Heather honey (that is honey from Ling heather) is thixotropic. In other words it is a sort of gel that flows only when agitated. There is elaborate (and expensive) extraction equipment that does just this but this is beyond the means of the hobby beekeeper. Small scale extraction of heather honey is normally done pressing crushed comb in a bag using a honey press. If the supers that are put on the hive when they go to heather contain frames with unwired foundation (or just starter strips) the honey-bearing comb can simply be cut out of the frame and crushed. Another way of marketing heather honey is as cut comb but there will inevitably be waste comb (offcuts) that contains honey that can only be extracted in a press.

The honey from Bell heather (so-called port wine honey because of its colour) is not thixotropic and can be extracted in the normal way. Bell heather comes into flower before Ling (mid-late July) and the honey often gets mixed with that from Ling.

Preliminary Filtering of the Honey

Most modern extractors have only a minimal reservoir under the cage for honey so the tap at the base has either to be left open or strategically opened to avoid flooding the bottom bearing with honey (which does not help free rotation!). We drain our honey through a conventional double strainer into 30lb plastic buckets and this removes all but the smallest debris from the honey. When handling honey bear in mind that the most silent thing in the world is overflowing honey! When the tap on the extractor is open the receiving container must be securely positioned under it and must have sufficient capacity to accommodate the amount to be delivered. There is nothing worse than finding a pool of your precious honey on the floor!

What to do with Wet Supers

These should be returned to the hives for cleaning and placed above the cover board (with ventilation holes open of course). It is good practice to clean boxes in the same apiary from which they came as this reduces the possible spread of diseases. If brood disease is a current or recent problem then boxes should be returned to the exact hive from which they were removed. The 'cleaner' boxes should only be put on the hives around dusk preferably on a cool evening. Whereas feeding sugar syrup presents few problems, the smell of exposed honey drives bees into a frenzy of robbing if you are not careful. Two days is quite long enough and if the 'cleaner' boxes are removed early in the morning (before the bees have started flying) they will usually be completely free of bees.

Some beekeepers advocate storing combs wet to avoid damage by wax moth but I have no experience of this practice. The boxes would need to be very securely sealed to avoid the attention of bees and wasps. The combs can look pretty awful when they are brought out of storage next spring but the bees will quickly clean them for re-use.

Preparing Honey for Bottling

A beekeeper with a few hives and only enough honey to supply family and friends may wish to proceed straight from extraction to bottling. In this case the honey can be transferred directly from the extractor to a (clean) ripener. It should be left standing for a few days for air bubbles and small fragment of wax to rise to the surface and then bottled.

Bulk honey is best stored in well-sealed, large containers (the 30lb bucket is a convenient size) until it is to be bottled. These buckets should be stored in a cool place that does not experience large fluctuations of temperature. Selling honey, particularly through a retailer, demands a higher standard of preparation than for home consumption and before it is bottled it needs to be fine filtered. We do not wish to remove pollen from our honey as this is a natural constituent and valued by most customers who purchase locally produced honey. So by a fine filter I mean about 400 microns (200 microns if you are really fussy).

Small quantities of honey can be dealt with using a filter bag or cloth. Some ripeners come with a filter/sieve that sits on top of them. They are usually fairly coarse but can be used in combination with a filter cloth. The best bit of kit for filtering honey, the 'Strainaway', is sadly no longer on the market. It worked using the suck from a vacuum cleaner to speed-up the process. It worked really well and we used one for about 15 years but unfortunately it was made of plastic which over time became brittle and eventually cracked (terminally). We now have a DIY stainless steel version made from a 50kg ripener.

Most British honey granulates in storage so for final filtration before bottling it has to be liquefied (ie. heated until it becomes clear again). This requires a temperature of 35-40°C but, as I said above, this needs to be done extremely carefully to avoid damage to the honey. All beekeepers who process any significant amount of honey should have a thermostatically heated warming cabinet which will accommodate at least one and preferably two 30lb tubs of honey. An old refrigerator cabinet (the 'larder' type is best suited) makes an excellent insulated warming cabinet. One can usually be obtained for free from an electrical equipment retailer or a re-cycling centre. First you need to strip out the cooling panel, pipe-work and motor and the only problem here is the refrigerant gas. The cooling system of a 'fridge should be decommissioned and the refrigerant re-cycled (not discharged to the atmosphere because it is a powerful greenhouse gas). So you really need to get hold of a refrigerator that has undergone this process or one that has failed by losing its refrigerant gas (which is a common failure mode for 'fridges). If the motor on the 'fridge works but there is no cooling effect then the likelihood is that the refrigerant has already leaked away.

Having obtained a suitable cabinet all you need now is a heater and a thermostat. Tubular (or bar) heaters are readily available and the 60W size is what you want (cost £15-£18). Make sure you get a model with two clips so that it can be attached to a block of wood on which it will stand in the bottom of the cabinet. The metal surface of these bar heaters does not get hot enough to present a fire hazard, melt plastic or burn you (a tungsten filament light bulb can do all these things!). Cheap thermostats which cover the range of temperature required (25-40°C) are more difficult to find as most central heating thermostats have a maximum temperature of 35°C. The one we have used (purchased about 12 years ago) is still available from RS (Radiospares) at a cost of just over £20 (RS Stock No. 375-0168) and covers the range 10-60°C. There is a circuit diagram printed on the inside of the case but if you have any doubts about how it should be correctly wired seek help from someone who knows about these things.

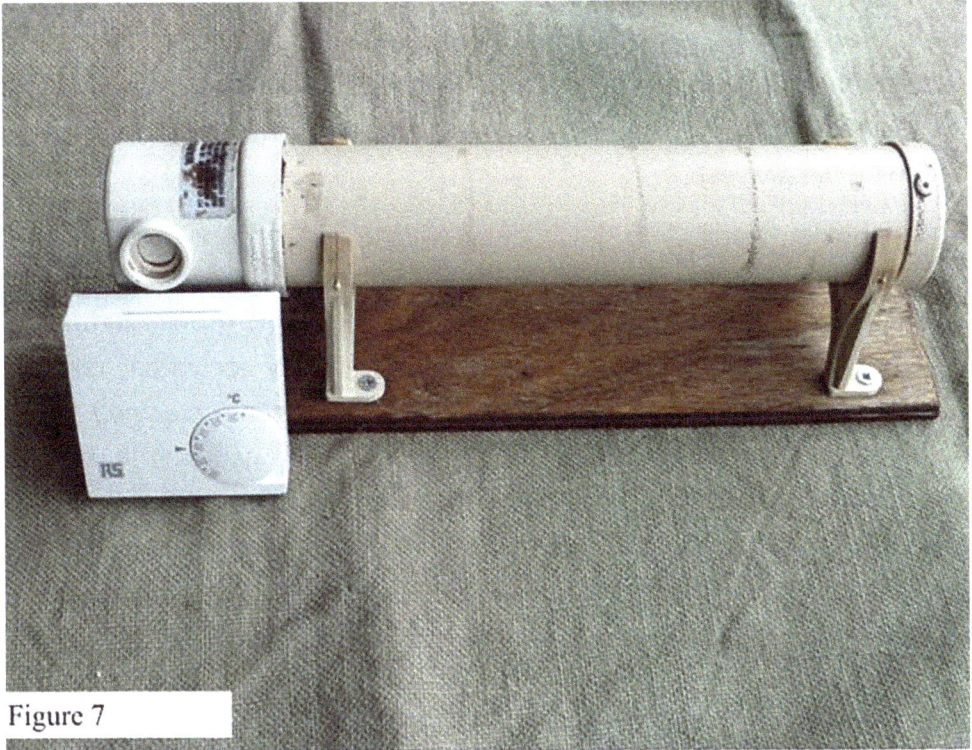

Figure 7

Set at a temperature of 40°C we find it takes 2½-3 days (in winter) to liquefy set honey. A temperature of 30-35°C will warm liquid honey sufficiently for filtering in about 24 hours. In this impatient age this may seem like a rather slow process but it means that at no time is any part of the honey heated to a temperature higher than that set on the thermostat. Warm air is also a gentler source of heat than water (lower specific heat) and there is time for heat transfer by conduction to occur.

Creamed or Soft-set Honey

Creamed honey is finely set honey that is meant to remain spreadable at room temperature throughout its shelf-life. Because the honey we harvest is so variable, the production of creamed honey to achieve a consistent result is something of a black-art. The basic principle is that you mix a small amount (5-10%) of finely granulated 'seed' honey with warmed (about 25°C) clear honey in a ripener and then stir it daily (or twice daily) for several days – or until you get fed-up. The stirring is best done at the lower temperature of 14-16°C at which the process of crystallization is most rapid. When you are satisfied that the mix has been sufficiently stirred it should be bottled before it becomes too stiff to flow. It may need gentle re-warming at this stage to speed up the process.

The main problem with making creamed honey is where does the 'seed' honey come from? Sometimes finely granulated honey occurs naturally but rarely to order. The sensible thing to do is to save some creamed honey from your previous batch to provide the 'seed' for the next one (if you can remember). An alternative solution is to make your own; warm some coarsely granulated honey to about 250C and cut (comminute) the crystals in a kitchen food processor (or blender) – it need 2 minutes or more to a achieve the right consistency.

Marketing Clear Honey

Some consumers express a strong preference for clear honey. Most British honey (apart from that derived from borage) will usually set in the bottle at sometime during its shelf-life; it may take as little as a fortnight or it may take several months. It may granulate unevenly, with large crystals slumped at the bottom of the jars. This affects the aesthetics rather than quality of the honey and can be rectified by re-warming. It is sometimes recommended that long-life clear honey should be produced by heating it to 60-65°C. This will give a product that will granulate for a long time, however this level heating produces pasteurized honey which is what you get with most supermarket honey. Strictly speaking it should only be sold a baking honey. In my view beekeeper honey should never be subjected with sort of abuse. After-all if they so desire, the customer can re-liquefy honey by placing the jar in warm water or by heating in a microwave using the defrost setting

The Honey Ripener and Bottling

By the time the honey is in the ripener all the processing of the honey and the potential to do it damage is over and it is just a matter of getting it into jars. Experience has shown us that the best temperature for filling jars is around 25°C. At a lower temperature the honey is very viscous and entrains air bubbles as it pours and at a higher temperature the same thing tends to happen.

Figure 8

The honey jars that you purchase have usually been re-packed since they came from the manufacturer. If you are in any doubt about their cleanliness they should be washed before use and this is where a dishwasher comes in handy. It is also good practice to warm jars in an oven before filling them and a temperature of about 50oC is suitable. Lids can also be a problem. Metal lids should not be re-used (for honey that is for sale) because once the lacquer on the threads has been damaged the acidity of the honey quickly corrodes the steel and an unsightly black deposit results. When fitting lids on jars it is easy to overlook specks of dirt and they need to be checked carefully before they are used. This may represent a shocking break with tradition, but we find the plastic lids for the 1lb honey jars are a great improvement over the metal ones; no corrosion, no sticking and they can usually be re-used.

Batch Numbers and Best Before Dates

In this modern world all food products have to be traceable to source and if you sell honey (and particularly through a retailer) you may be asked by Food Standards to show that you keep proper records and comply with the labelling regulations. This is not difficult; each ripener of honey that you bottle should be given an individual Batch number. You should keep a record of where it came from and when it was bottled and, if possible, the water content. The Batch number should also be written somewhere on the labelling of the jar and a record kept of to whom all bulk sales were made. Nothing silly is required and a few jars sold direct to the customer need not be recorded in this way. Additional advice on the legalities of handling and labelling honey can be found in the form of a pdf on www.beekeeping.org.uk/food_safety.pdf or look at www.food.gov.uk/wales/labelwales/labelwales

Water content is the main determinant of Best Before date and we keep ours within the range of 9-24 months. The date is handwritten of the tamper-evident label using a marker pen. It is not legal to sell honey with a water content greater than 20% (21% for heather honey). There are other legal requirements for the wording that appears on labels but I am not an expert on that subject but there have been articles in beekeeping magazines giving the letter of the law. The main suppliers of labels can usually be relied on to provide a legal product.

How to be Kind to Your Honey

To avoid degrading the quality of the honey that you have taken off the hive and getting it into the jar in the best possible condition here are some of the key do's and don'ts (most of which have already been covered in more detail above).

- Do not leave boxes of combs standing about in an environment in which they can pick up water – extract immediately or store where the relative humidity is 60% or less.

- Do not overheat the honey in the combs. A temperature of 25°C is ideal for extraction – the honey flows easily and the comb is not significantly weakened and liable to collapse.

- Do not use a heated uncapping knife as this will caramelise the honey it touches and result in tainting.

- Do not use a heated uncapping tray unless it is turned down really low – 40oC should be the maximum temperature.

- On no account use the old trick of separating the honey from the cappings by melting the wax as this requires a temperature of 64°C+. Honey that has been over-heated in this way should only be sold as 'cooking' or 'baking' honey and should be clearly labelled as such.

- It is impossible to avoid some loss of volatiles during extraction, when the honey is spraying out of the cells in droplet form, but a lid on the extractor may help a bit (and also save possible injury to stray fingers).

- Unless you only have a small amount of honey that will be bottled almost immediately, store honey in bulk in well-sealed tubs or buckets – 30lbs (13.6kg) is a convenient size.

- Store honey in a cool place with a fairly stable temperature and preferably in the dark.

- When re-heating bulk honey for bottling do it slowly and in a thermostatically controlled cabinet – the traditional light bulb powered honey warmer can lead to over-heating.

- An ideal temperature for ripening and bottling honey is about 25°C – it doesn't damage the honey and seems to give a nice bubble-free flow when pouring.

Figure 9

In Conclusion

Honey is a wonderful and complex product that is easily damaged during its journey from the hive to the jar. It is up to us to bring it to market in the best possible condition so that we maintain reputation and premium status of locally produced, beekeeper honey.